DUNNIGAN
I N D U S T R I E S ,   I N C .
Proven Resources for Leadership Development

RUNNING WITH MODERN MILITARY CADENCE

Produced and distributed by:

DUNNIGAN Industries, Inc.
P.O. Box 1387
Fortson, Georgia   31808-1387

Manufactured in the United States of America
10 9 8 7 6 5 4

INTRODUCTION

As mentioned in volume one, the military is ever changing. With the abundance of cadence material available, we want to move forward with the changing times and keep service members up to date. Publishing many of the, "*oldies, but goodies*" remains part of our tradition here at Dunnigan Industries. Our goal is to continue to include tradition while adding new material building on the uninterrupted supply of military heritage in rhythm.

Because this book is complied from many sources, the cadences are in no particular order and many are universal and require no branch identification. Don't get caught short the next time you're called out during a run to sing cadence. Always remember to begin when the left foot hits the street and you're well on your way to **Running with Modern Military Cadence**.

TABLE OF CONTENTS

Dunnigan Industries, Inc.

SUPERMAN

I don't know, but I think I can
Take the S from Superman

Early this morning, late last night
Me and Superman had a fight

I hit 'em with my left, I hit 'em with my
right
I hit 'em in the heat with some cryptonite

I hit 'em so hard I busted his brain
Now I'm dating Lois Lane

Jimmy Olsen jumped me from the rear
I kicked 'em in the knee and bit 'em on the
ear

I kicked 'em with my foot and hit 'em with
my hand
Now he's buried with Superman

BATMAN

I don't know, but I think I might
Look for Batman Saturday night

I don't know, but I think I will
Take a little ride in the bat mobile

Me and Batman we fought too
I kicked 'em in the head with the heel of my shoe

I kicked 'em with the toe, kicked 'em with the heel
Now I'm kicking in the bat mobile

ALL THE RIGHT STUFF

Mamma, mamma don't you cry
Your baby little boy ain't gonna die

I'll make him tough, I'll make him strong
You won't know 'em when he gets home

He'll be strong and he'll be tough
He'll be made of all the right stuff

Loyalty, duty and respect
A lot of selfless service you can bet

And when he speaks he'll be heard
Honor and integrity in his words

leadership it is the key
It's the backbone of the Army

Dunnigan Industries, Inc.

ALMIGHTY INFANTRY

I am the almighty Infantry
Pick up you weapons and follow me

I hump all day and I hump all night
On my way to the firefight

We're trained to kill we're trained to fight
We do it all day, we do it all night

This is my family, this is my team
U.S. Army fighting machine

Together we'll conquer the enemy
Cause we're the almighty infantry

Contributed by,
PVT Jonathan Thomas
PVT Thomas Reid

KEEP IT TIGHT

Mamma, mamma don't you cry
Your little boy ain't gonna die

1 push up, 2 push ups, 3 push ups, 4
My best friend is the concrete floor

Drill Sergeant runs us all day and night
His favorite line is, "keep it tight"

Keep it tight
Dress it right
Cover down

One hundred and ten is what we give
That's the only way we're gonna live

Cause I'm hard core
Infantry

Contributed by,
PVT Thomas Borrowski
PV2 Colin Urbanski

RUNNING FOR MY HEALTH

Left, left, left right oh left
Your right left
Stay in step
Left, right, left
Running for my health

I love to double time
I do it all the time
It makes me feel fine
Clears up my mind

Left, left, left, right oh left
Your right left
Stay in step
Left, right, left
I'm running for my health

DOG TAGS

Rollin', rollin', rollin'
Ohhhh my feet are swollen

Don't let your dog tags dangle in the dirt
Pick up your dog tags and tuck 'em in your shirt

Rollin', rollin', rollin'
Ohhhh my knee is swollen

Don't let your dog tags dangle in the rocks
Pick up your dog tags and put 'em in your socks

Rollin', rollin', rollin'
Ohhhh my back is swollen

Don't let your dog tags dangle in the sand
Pick up your dog tags and hold 'em in your hand

Rollin', rollin', rollin'
Ohhhh my head is swollen

Don't let your dog tags dangle in the mud
Pick up your dog tags and hand 'em to your
bud

ROLLIN' ROLLIN' ROLLIN'

Rollin', rollin', rollin'
Oh my _____ are (is) swollen
(ankles, feet, legs, knees, etc.)

Don't let your dingle-dangle dangle in the mud
Pick up your dingle-dangle, give it to your bud

Don't let your dingle-dangle dangle in the dirt
Pick up your dingle-dangle, put it in your shirt

Don't let your dingle-dangle dangle on the ground
Pick up your dingle-dangle, shake it all around

Don't let your dingle-dangle dangle on the track
Pick up your dingle-dangle, put it in your pack

Dunnigan Industries, Inc.

DINGLE-DANGLE

Left, left, oh right left
Left, right, left
Keep it in step

But, don't let your dingle-dangle dangle in the dirt
Pick up your dingle-dangle, put it in your shirt

Left, left, oh right left
Left, right, left
Keep it in step

But, don't let your dingle-dangle dangle in the mud
Pick up your dingle-dangle, hand it to your bud

Left, left, oh right left
Left, right, left
Keep it in step

But, don't let your dingle-dangle dangle in the snow

Pick up your dingle-dangle, tie it in a bow

Left, left, oh right left
Left, right, left
Keep it in step

But, don't let your dingle-dangle dangle too low
Pick up your dingle-dangle, and let's go

Left, left, oh right left
Left, right, left
Keep it in step

MY TWO DOGS

I got dog, his name is Jack
If you throw him a stick, he won't bring it
back

He's got better things to do
You see, old Jack is Airborne too

Jack be nimble, Jack be hard
Not a dog for miles comes in my yard

Jack be quick, Jack be bad
Jack has got his Ranger tab

When he was little pup
We made him a chute and took him up

Standing in the door there was no doubt
He was wagging his tail when he went out

Cause he's A.I
R.B.
O.R.
N.E.
Airborne

Ranger
All the way

I got another dog, I named him Blue
Blue wanna be a Ranger too

So, early one day I took away his chow
And put some motivation in his bow wow

I made him walk for 15 days
And put old Blue into a zombie haze

Now my Blue is a Ranger too
Mess with him and he'll bite you

Cause he's A.I
R.B.
O.R.
N.E.
Airborne
Ranger
All the way

BLUE

Had an old dog who's name was Blue
Blue wants to go to SCUBA school

Bought him a tank and four little fins
And took him down where he got the bends

Same old dog who's name was Blue
He now want to go to Ranger school

Took him to the field, took away his chow
Put a little motivation in his bow wow

Still had the dog who's name was Blue
He now want's to go to Airborne school

Got him a chute, strapped it to his back
Now old Blue stands tall, looks strack

That Airborne dog who's name was Blue
Got his orders for Jungle school

Took him on down to Panama
And that's the last of Blue I ever saw

ONE MILE

Here we go
On the run
Just for fun

One mile
I can hang

Two miles
You can hang

Three miles
We can hang

Ha!
Ha, ha
Hooah
Ha, ha

Four miles
I can do it

Five miles
We can do it

Six miles
Nothing to it

Ha!
Ha, ha
Hooah
Ha, ha

Little run
To the sun

Ha!
Ha, ha
Hooah
Ha, ha

I can hang
Can you hang

I can hang
With the pain

ONE MILE

One mile
No sweat

Two miles
Better yet

Three miles
Gotta run

Four miles
To the sun

Dunnigan Industries, Inc.

HEY UP FRONT

Hey up front if you want to bump, say what

(chorus)	Don't stop the PT run
or	Don't wash your PT shorts
or	Napalm the CO's hooch

Hey in the middle if you want to wiggle, say what

(chorus)	Don't stop the PT run
or	Don't wash your PT shorts
or	Napalm the CO's hooch

Hey in the rear if you want some beer, say what

(chorus)	Don't stop the PT run
or	Don't wash your PT shorts
or	Napalm the CO's hooch

A little louder

(chorus)	Don't stop the PT run
or	Don't wash your PT shorts
or	Napalm the CO's hooch

CAN'T YOU SEE

CO, CO, can't you see
This little old run ain't nothing to me

I run all day, I run all night
If you didn't know it, I'm fit to fight

First Sergeant, First Sergeant can't you see
You're gonna lose a whole lot before you
lose me

I can run all day, I can run all night
I can run on down to a firefight

Dunnigan Industries, Inc.

OLD LADY AIRBORNE

I saw an old lady running down the street
Had a chute on her back, jump boots on
her feet

I said, hey old lady where you going to
She said U.S. Army Airborne school

What you gonna do when you get there
Jump from a plane and fall through the air

I said, hey old lady ain't you been told
Airborne school is for the brave and the
bold

She said, hey now soldier, don't be a fool
I'm an instructor at Airborne school

OLD LADY RANGER

I saw an old lady marching down the road
Had a knife in her hand and a 90 pound
load

I said hey old lady where you going to
She said U.S. Army Ranger school

What you gonna do when you get there
Jump, swim, and kill without a care

I said, hey old lady ain't you been told
Ranger school is for the brave and the bold

She said, hey young soldier don't be a fool
I'm the lead instructor at Ranger school

OLD LADY DIVER

I saw an old lady running 'round the track
Had fins on her feet and a tank on her back

I said, hey old lady where you going to
She said, U.S. Army SCUBA school

What you gonna do when you get there
Swim under water and never breath air

I said, hey old lady ain't you been told
SCUBA school is for the young and the bold

She said, hey now diver don't be a fool
I'm the head instructor at the SCUBA school

UP IN THE MORNING

I don't know, but I think I might
Jump from an airplane while in-flight

Soldier, soldier have you heard
I'm gonna jump from a big iron bird

Up in the morning in the drizzling rain
Packed my chute and boarded the plane

C-130 rollin' down the strip
Airborne Rangers on a one-way trip

Mission top secret, destination unknown
Don't even know if we're ever coming home

When my plane get up so high
Airborne troopers gonna dance in the sky

Stand up, hook up, shuffle to the door
Jump right out and count to four

If my main don't open wide
I got a reserve by my side

If that one should fail me to
Look out ground I'm a coming through

If I die on the old drop zone
Box me up and ship me home

Burry speakers all around my head
So I can rock with the Grateful Dead

Burry speakers all around my toes
So I can rock with Axel Rose

If I die on a Chinese hill
Take my watch or the commies will

If I die in the Korean mud
Burry me with a case of Bud

Put my wings upon my chest
And tell my mom I did my best

UP AND OUT OF THE RACK

Up in the morning and out of the rack
Grab my clothes and put them on my back

Running 'cross the desert in the shifting
sand
Drill sergeant look I'll give you a helping
hand

Up in the morning with a whistle and a yell
I know that voice and I know it well

Drill sergeant says you better hit the floor
And don't be walking going out the door

I like fun and I like wine
But all I do is double time

Double time here and double time there
Man this life, it's the best anywhere

GRANNY DOES PT

When my granny was 91
She did PT just for fun

When my granny was 92
She did PT better than you

When my granny was 93
She did PT better than me

When my granny was 94
She did PT more and more

When my granny was 95
She did PT to stay alive

When my granny was 96
She did PT just for kicks

When my granny was 97
She upped, she died, and went to heaven

When my granny was 98
She did PT at the pearly gates

When my granny was 99
She was in heaven doing double time

MY OLD GRANNY

When my old granny was 91
She joined the Army just for fun

When my old granny was 92
She did PT in combat boots

When my old granny was 93
She practiced PLFs from a tree

When my old granny was 94
She knocked out 10 and begged for more

When my old granny was 95
She fired expert and that's no jive

When my old granny was 96
She went Airborne just for kicks

When my old granny was 97
She up and died and went to heaven

She met St. Peter at the pearly gates
Said hey St. Peter I hope I ain't late

St. Peter looked at granny with a big ol' grin
Said get down granny and knock out ten

She knocked out ten and did ten more
Dedicated them to the NCO corps

Peter looked at granny and said you're so
cool
We're sending you back for Ranger school

Granny said to Peter, hey I ain't no fool
I could be teaching at that dog gone school

A.I.R.B.O.R.N.E.

A- is for Airborne

I- is for In the sky

R- is for Ranger

B- is for Bonafied

O- is for On the go

R- is for Rock and Roll

N- is for Never quit

E- is for Everyday

Cause I'm Airborne all the way
Super-duper paratrooper

R.A.N.G.E.R.

R- is for Rough and tough

A- is for All the way

N- is for Never quit

E- is for Excellence

R- is for Ranger

I want to be an Airborne Ranger
Live the life blood, guts, and danger
I want to cut off all of my hair

RAVING MAD

Airborne Ranger raving mad

He's got a tab I wish I had

Black and gold and a half moon shape

Airborne Ranger he's gone ape

Jumpin' through windows, kicking down the walls

Airborne Ranger's having a ball

So it there's trouble in the world today

Call on the men in the tan berets

WHEN I GET TO HEAVEN

When I get to heaven
Saint Peter's gonna say

How'd you earn your living
How'd you earn your pay

And I'll reply with a little bit of anger
Made my living as an Airborne Ranger

Airborne Ranger
Ranger danger
Airborne Ranger
Tan beret danger

I love to double time
I do it all the time

Dunnigan Industries, Inc.

EARN YOUR LIVING

When I get to heaven
St. Peter's gonna say

How'd you earn your living boy
How'd you earn your pay

I'll reply with a whole lot of anger
Made my living as an Airborne Ranger

Blood and guts and a whole lot of danger
That's the life of an Airborne Ranger

When I get to hell
Satan's gonna say

How'd you earn your living boy
How'd you earn your pay

I'll reply with a fist to his face
Made my living laying souls to waste

C - 130

C-130 rollin down the strip
Airborne daddy on a one way trip

Mission uncertain, destination unknown
We don't know if we're ever coming home

Stand up, hook up, shuffle to the door
Jump right out and count to four

If my main don't open wide
I got a reserve by my side

If that one should fail me too
Look out ground I'm coming through

Slip to the right and slip to the left
Slip on down, do a PLF

Hit the drop zone with my feet apart
Legs in my stomach, feet in my heart

BURY ME

If I die on the old drop zone
Box me up and ship me home

Pin my wings upon my chest
Bury me in the leaning rest

If I die in the Spanish Moors
Bury me deep with a case of Coors

If I die in Korean mud
Bury me deep with a case of Bud

If I die in a firefight
Bury me deep with a case of Lite

If I die in a German blitz
Bury me deep with a case of Schlitz

If I die, don't bring me back
Just bury me with a case of Jack

CHAIRBORNE RANGER

It's one thirty now on the strip
Cairborne daddy gonna take a little trip

Stand up, lock up, shuffle to the door
The club for lunch and home by four

If there's something to decide
Close your door and try to hide

Every time you get a call
You're out playing racquetball

First revise the SOP
Make a change in policy

Ours is not to wonder why
It's written down in the LOI

God forbid we go to war
All that paperwork would be a bore

Let me stay behind my desk
Anything is better than the leaning rest

Chairborne Ranger, that's what I am
One of a kind, I'm an AG man

COON SKIN

Coon skin and alligator hide
Make a pair of jump boots just the right
size

Shine 'em up, lace 'em up, put 'em on your
feet
A good pair of jump boots can't be beat

Birdy, birdy in the sky
Dropped some white wash in my eye

Ain't no sissy, I won't cry
I'm just glad that cows don't fly

NOTHING TO DO

AG, AG, who are you
TDA with nothing to do

Go to PT at 9 am
Then to the pool to have a swim

Racquetball from 9 to 10
Recover with a tonic and gin

Lunch from 11 until noon
Your day will be over soon

Volleyball from noon 'til 3
Keep really busy, can't you see

Now it's 4, your day is through
I wish I was AG too

AMEN

Aaaaaaaamen (chorus sung continuously by all)

(Caller sings between chorus lines)

Sing it over

Sing it louder now

Sing it soft now

Real loud now

Real soft now

Hallelujah

Praise the Lord now

G.I.

G.I. coat and G.I. comb
Gee, I wish that I were home

G.I. coat and G.I. gravy
Gee, I wish I'd joined the Navy

WHER'VE YOU BEEN

AWOL, AWOL wher've been
Down in the bar, drinking jin

What you gonna do when you get back
Sweat it all out on the PT track

ENGINEERS ARE NUMBER ONE

Engineers are number one
They call us when there's work to be done

Aviation is all we hear
We do the work, they drink the beer

Pistol ranges, soccer goals
Road extensions, we do it all

ENGINEER, ENGINEER

Engineer, Engineer running down the road
Running so fast makes the others look old

We are running hard and we're running long
Still singing another stupid song

Build a road or cut down a tree
Or dig some graves for the Infantry

Working hard and working all day
Knocking down anything that gets in the way

DESERT SAND

Went down to see the man
He gave me orders for the desert sand

I packed my weapon, packed up my ruck
They threw me in this 5 ton truck

As I look out with a glassy glare
The next thing I know I'm in the air

When we land, it's dark and warm
They tell me I'm at the Desert Storm

For the next six months, this is your home
No running water, no telephone

Sadam Hussein he said to me
I want to be all I can be

I'll pack your weapon, I'll pack your ruck
As for Iraq they have no luck

FORCE RECON

Pint my face black and green
You won't see me I'm a recon Marine

I slip and slither into the night
You won't see me 'til I'm ready to fight

You'll run in the bushes, you'll try to hide
But that's where I live, you're sure to die

You won't see me 'til it's too late
A flash of my bang will be your fate

GEORGE S. PATTON

In 1934 we took a little trip
Me and George S. Patton headed down to
Mississipp'

We shot our main guns 'til the barrels
melted down
Then we grabbed a couple legs and we
went a couple rounds

Cause we're mentally able and we're
physically fit
And if you ain't Armor, you ain't it

GRANDDADDY

My Granddaddy was a horse Marine
When he was born, he was wearing green

Ate his steak six inches thick
Picked his teeth with a swagger stick

Drinking and fighting and running all day
Granddady knew no other way

Lived everyday of his life for the corps
So they sent him off to war

Went to the islands to fight the Japanese
Caught some schrapnel in the knees

Later, at Chosen Reservoir
Caught a bullet in his derriere

Went to a country called Viet Nam
To fight some people called Viet Cong

Found himself in a firefight
Came back home on a Medevac flight

Now Granddaddy just sits there
Marking his time in his rocking chair

IRAQI BLUES

Send the troops before it's too late
Saddam has invaded Kuwait

Grab your rifle and get a tan
You can scratch a rotation plan

President Bush was talking tough
We didn't know it would get that rough

Thought Saddam was a man of reason
Now we've got him for the crime of treason

America's become divided as such
They don't like that war stuff much

Cussing and a picketing that's the scoop
Throw rocks at me but you support out troops

People are starting to understand
Saddam Hussein's one crazy man

Gasses his people and tortures them too
Saddam this cluster bomb's for you

Burning oil and acid rain
SCUD missile, desert terrain

Shipped my butt straight overseas
Hey, who cut down all the trees

One - two - three and four
Sometimes to get peace, you gotta make some war

If we don't nuke 'em 'til they glow
We'll die for more than Texaco

Stormin' Norman made a plan
Now we're gonna kill who's in command

When we're through kicking his butt
We'll pay for gas and it won't cost much

This is my story and it is true
I call this song the Iraqi Blues

Saddam, act stupid and I won't refuse
To put you on the 10 o'clock news

JUMP INTO BATTLE

Jump, jump, jump into battle
Hear those sixty's rattle

Shoot, move and cover my brother
Write a letter to my mother

Jump, jump, jump into battle
Hear those fifty cals rattle

Shoot, move and cover my brother
Write a letter to my mother

MARINE BY GOD

Born in the woods, raised by a bear
I got a double set of jaw teeth and a triple
coat of hair

Two brass balls and a cast iron rod
I'm a mean devil dog, a Marine by God

1775

Back in 1775
My Marine Corps came alive

First there came the color blue
To show the world that we are true

Next there came the color red
To show the world the blood we shed

Finally there came the color green
To show the world that we are mean

TALE OF THE RECON MARINE

Way, way back in the dawn of time
In the valley of death were the sun don't
shine

A mighty fighting man was made
From an M-16 and a live grenade

He looked mighty big with his ALICE pack
He drove mighty mean with his Cadillac

This mighty fighting lean green machine
Goes by the title, Recon Marine

Roll on your left and roll on your right
Roll on your left we love to double time

TARZAN AND JANE

Tarzan and Jane were swinging from a vine
Sippin' from a bottle of whisky double wine

Jane missed the vine and then she fell
When she hit the ground, she gave a little yell

Ai - e - ai
Mmm mmm
Feels good
Ai - e - ai
Mmm mmm
Real good

Tarzan and Cheetah were swinging from a vine
Sippin' from a bottle of whisky double wine

Cheetah missed the vine and then he fell
When he hit the ground, he gave a little yell

Ai - e - ai
Mmm mmm
Feels good

Dunnigan Industries, Inc.

Ai - e - ai
Mmm mmm
Real good

TERRIBLE JAM IN VIET NAM

Come on all of you big strong men
Uncle Sam needs your help again

Got himself in a terrible jam
Way down yonder in Viet Nam

So put down your books and pick up a gun
We are all gonna have a whole lotta fun

Come on Wall Street don't be slow
Man this is war so go, go, go

There is a lot of good money to be made
Supplying the Army with the tools of the
trade

Just hope and pray that if we drop the
bomb
We go and drop it on Viet Nam

Come on generals, let's move fast
Your big chance is here at last

Now we can go out and get those reds

Cause the only good commie is one that's dead

UP FROM A SUB

Out in the sky in the middle of the night
When we hit the deck we're ready to fight

Up from a sub 60 feet below
We SCUBA to the surface and we're ready
to go

We're gonna back stroke, side stroke, swim
to shore
When we hit the beach, we're ready for war

Singing hoo-ya, hoo-ya, hey
Hoo-ya, running day
Singing hoo-ya, hoo-ya, hey
Just another PT day

Well chief caught a round right between the
eyes
And corpsman thought for sure the chief
would die

But chief stood up straight as any man
And killed four commies hand-to-hand

Well 20 seconds later there was not a sound
And 50 dead commies were lying around

Singing hoo-ya, hoo-ya, hey
Hoo-ya, running day
Singing hoo-ya, hoo-ya, hey
Just another PT day

Now superman may be the man of steel
But he ain't no match for a Navy Seal

Now chief and superman they got in a fight
Chief hit him in the head with some
cryptonite

Superman fell down on his knees in pain
Now chief is dating Lois Lane

Singing hoo-ya, hoo-ya, hey
Hoo-ya, running day
Singing hoo-ya, hoo-ya, hey
Just another PT day

SCUBA BLUE

Well I've got a dog and his name is Blue
And Blue wants to be a Navy Seal too

So, I bought him a mask and four little fins
I took him to the ocean and threw his butt in

Blue came back and to my surprise
With a shark in his teeth and a gleam in his eyes

WHEN I DIE

When I die, burry me deep
With two crossed rifles laid beneath my feet

By my side a forty five I wear
And don't forget to pack my PT gear

Because early one morning around zero five
The ground's gonna shake, they'll be thun-
der in the sky

Don't you get alarmed, don't you come
undone
It's just me a Chesty Puller on a PT run

FILLERS

Yea, I'm hard-core

Lean and mean

Fit to fight

Out of sight

One mile

No sweat

Two miles

Better yet

Three miles

I can make it

You can make it

Huah

 Dunnigan Industries, Inc.

A - ha

Huah

A - ha

Hard-core

Lean and mean

On the scene

Army green

UH-60

UH-60 flying high
These are the whimps that snivel and cry

Stand up, snap in, slide down a rope
An Air Assault soldier ain't nothing but a
joke

CAPTAIN'S BARS

Twinkle, twinkle, little star
Where did you get those captain's bars

Please don't tell me, let me guess
Two box tops and OCS

FIRE MISSION

Fire mission, fire mission, coming down
81mm on the ground

FDC this is OP right
I have the enemy in my sights

FDC this is OP left
Send me some of that silent death

OP left this is FDC
I have HDP and Willie Pete

Boom
Boom, boom

AIRBORNE PATHFINDER

Up from the desert of Iraq
An Airborne Pathfinder dawns his pack

I walk all day, set up an LZ at night
Airborne Pathfinder is out of sight

Cobra one this is Pathfinder three
I have a rag-head a pestering me

Out of the sky through the darkness of
night
40 mike-mike rockets are in flight

Well drop five zero, 100 to the left
Rounds so close that I scared myself

It was prettier than a rainbow, horrified
stealth
40 mike-mike rounds were raining down
death

Cobra one, this is Pathfinder three
A little bit closer, I'm almost free

Well, I threw two grenades and I rolled to my left
It didn't bother me that I was facing death

I fired to my left, and I fired to my right
Heads exploding, it was a hell of a sight

Raider two, this is Pathfinder three
I'm on the run to my new LZ

Signal out, can you identify
Green smoke rising in the sky

Touch down, take off in six seconds flat
Special operations, what you think of that

Flying back home, swelled up with pride
I'm gonna find a country girl and make her my bride

Contributed by,
SFC Ricky Johnson

Dunnigan Industries, Inc.

MARINE CORPS

My Corps

Your Corps

Our Corps

Marine Corps

Is hard-core

SHOUT MARINE CORPS

C-130 rolling down the strip
Airborne daddy gonna take a little trip

Stand up, hook up, shuffle to the door
Jump on out and shout Marine Corps

Dunnigan Industries, Inc.

MAMMA TOLD JOHNNY

Mamma told Johnny not to go down town
Marine recruiter was hanging around

Johnny went down town anyway
To hear what he had to say

Recruiter asked Johnny what he wanted to
be
Johnny said he wanted to join the Infantry

Next Johnny woke in the pouring rain
Staff sergeant said it was time for pain

They jogged five miles then ran for three
This is the life in the Infantry

Contributed by,
CDT PFC Randall Whiehouse

THE HAMILTON SHAKEDOWN

We are ready to fight, we're ready to drill
We are ready to do the DI's will

We are ready to lead, we're willing to follow
We'll get our butts outta the hollow

They say to err is human, to forgive is
divine
That ain't no part of that Corps of mine

Gimme a hoorah
Gimme a hoorah
I said gimme a hoorah
Well allright right

Contributed by,
Linda Cimino

WE ARE AIRBORNE

Hey, hey
Hey, hey

We are Airborne
Mighty, mighty Rangers
We can fly higher than Superman

We are Airborne
Mighty, mighty Rangers
We are stronger than all your men

We are Airborne
Mighty, mighty Rangers
We can swim faster than Aquaman

We are Airborne
Mighty, mighty Rangers
We can climb faster than Spiderman

We are Airborne
Mighty, mighty Rangers

Contributed by,
PVT Steven Nelson

AIRBORNE RANGER

Send my regiment off the war, hey

C-130 rolling down the strip
64 Rangers on a one-way strip

Mission unspoken, destination unknown
We don't know if we're ever coming home

Stand up, hook up, shuffle to the door
Jump right out and count to four

If my main don't open wide
I got a reserve on my side

And if that one should fail me too
Look out below Airborne Ranger coming
through

Hit the ground in the middle of the night
Spring right up into a firefight

Contributed by,
PVT Steven Nelson

Dunnigan Industries, Inc.

GOING FOR THE GOLD

I don't know if you've been told
But what I want is to get that gold

One, two, three, four hey
Somebody, anybody, start a war, hey

To pin that gold bar above my chest
Will make me a part of America's best

One, two, three, four hey
Somebody, anybody, start a war, hey

OCS is mighty rough
That's what makes candidates tough

One, two, three, four hey
Somebody, anybody, start a war, hey

When we graduate from OCS
We'll represent Pennsylvania's best

One, two, three, four hey
Somebody, anybody, start a war, hey

Every day we run PT
They run the sweat right out of me

One, two, three, four hey
Somebody, anybody, start a war, hey

Hard-core training makes us mean
The meanest things you've ever seen

One, two, three, four hey
Somebody, anybody, start a war, hey

Combat ready every day
That's how we earn out monthly pay

One, two, three, four hey
Somebody, anybody, start a war, hey

Fighting men go where they're sent
Blood and guts help pay the rent

Contributed by,
Fidel Gonzalez, Jr.

PAPPY BOYINGTON

Pappy Boyington ruled the air
The Japanese thought he fought unfair

Twenty-eight zeros he shot down
Pappy was the best fighter pilot around

A Medal of Honor received with grace
Pappy was the Marine Corps number 1 ace

UP IN THE MORNING

Up in the morning half passed three
First Maunakea bringing heat

NCOs all around his desk
Had a second lieutenant in the leaning rest

First Sergeant, First Sergeant, can't you see
You can't bring no smoke on me

I can run to Maryland just like this
All the way to Baltimore and never quit

I can run to Florida just like this
All the way to Miami and never quit

I can run to Washington just like this
All the way to DC just like this

I can run to Alabama just like this
All the way to Anniston and never quit

I can run to Fort Bragg just like this
All the way to Carolina and never quit

Dunnigan Industries, Inc.

KAMIKAZE KILLERS

Hey, all the way
We run every day

Airborne Rangers wear the tan beret
Kamikaze killers and we earn out pay

Jumping out of airplanes, running through
the swamp
Uncle Sam gets in trouble, Rangers gonna
stomp

Our minds are like computers, our fists are
like steel
If one don't get you then the other one will

HIT THE BEACH

Up from a sub 50 feet below
Up swims a man with a tab of gold

Back stroke, side stroke, heading for shore
He hits the beach and he's ready for war

A grease gun, K-bar by his side
These are the tools that he lives by

How to kill a search team, a hostage snatch
Out of the sub and back into the hatch

Hand-to-hand combat behind the line
Airborne Ranger just a killing time

He jumps through windows and kicks down walls
Airborne Ranger just having a ball

10,000 NASTY LEGS

One there was ten thousand nasty legs
Caught in the valley by the NVA

The President was worried, he wondered
what to do
So he called on me and he called on you

We jumped on in and did the job well
Sent those commies straight to hell

So if you have a problem, don't delay
Call on the men with the tan berets

Singing hey all the way
Rangers lead the way

One, two, three, four hey
Every night we pray for war hey

One, two, three, four hey
Somebody, anybody, start a war hey

Left, right, left, right, left, right, kill
Left, right, left, right, you know I will

SCUBA DOG

I had a dog his name was Blue
Blue wanted to be a SCUBA diver too

So I bought him a mask and four little fins
And took him on down until he got the
bends

Blue recovered and to my surprise
He now dives to one thirty five

RAVING MAD

Airborne Ranger raving mad
He got a tab I wish I had

Black and gold in a half moon shape
Airborne Ranger, he's gone ape

AIRBORNE (TO AMEN)

Here we go now

(Chorus)
Airborne

All together now

(Chorus)
Airborne

One more time now

(Chorus)
Airborne, Airborne, Airborne

All together now

(Chorus)
Airborne

Feeling good now

(Chorus)
Airborne

All the way now

(Chorus)
Airborne, Airborne, Airborne

SITTING ON A MOUNTAIN TOP

Sitting on a mountain top beating a drum
Beat so hard that the MPs come

MP, MP, don't arrest me
Arrest the grunt behind the tree

He stole the whisky, he stole the wine
All I ever do is double time

Cause I'm hard-core
Mean and lean

Cut clean
Looking good

Aught to be
Hollywood

Every day
One mile
No sweat

Two miles
Better yet

One, two, three, four hey
One, two, three, four hey

MY RECRUITER

My recruiter told me a lie
You're gonna be Airborne, jump and fly

So I signed my name on the dotted line
Now, all I ever do is double time

Up the hill
Down the hill

Through the hill
Airborne

RANGER

One, two, three, four, hey
One, two, three, four, hey

Here we go
Everyday

Gotta run
Airborne

Gotta be
Number one

Airborne
Ranger

R - running hard
A - all the way
N - never quit
G - gotta go
E - every day
R - Ranger

Hard core
Super trooper

Hard core
Paratrooper

Dunnigan Industries, Inc.

JUMP, SWIM, KILL

Saw an old lady running down the street
She had a chute on her back and boots on
her feet

I said old lady where you going to
She said U.S. Army Airborne school

I said hey old lady you're too darn old
Leave that training to the young and the
bold

She said hey young punk who you talking
to
I'm an instructor at the Airborne school

I saw an old man coming down the track
He had fins on his feet and tanks on his
back

I said hey old man where you going to
He said U.S. Army SCUBA school

Whatcha gonna do when you get there
Swim under water and never breathe air

I saw a young man coming down the road
He had a knife in his hand and a 90 pound
load

I said hey young man where you going to
He said U.S. Army Ranger school

Whatcha gonna do when you get there
Jump, swim, kill without a care

MOMMA AND PAPA

Momma and papa were lying in bed
Papa rolled over to Momma and said

Give me some
PT

Good for you
Good for me

Get up in the morning in the morning with
the rising sun
We run all day until the work is done

One mile
Having fun

Mile two
Good for you

Mile three
Good for me

Mile four
Want some more

Mile five
Keep it alive

Mile six
Great kicks

Mile seven
This is heaven

Mile eight
This is great

Mile nine
Mighty fine

Mile ten
Do it again

Dunnigan Industries, Inc.

ARMY VS. MARINES

I don't know what I've been told
The Marine Corps thinks it's mighty bold

They don't know what the Army can do
We are proud of our history too

Out looks and style may not be smooth
But you oughtta see this Army move

Look to your left and what do you see
A bunch of jar heads just looking at me

Shout it out and sing it loud
I am a soldier and I'm mighty proud

GROWING UP IN THE BAD LANDS

When I was young and growing up in the
bad lands
Momma said son what do you wanna be
when you're a man

Something with guts and a whole lot of
anger
Son you wanna be an Airborne Ranger

When I was in school my teacher said to
me
Now that you're done son whatcha wanna
be

Something with guts and a whole lot of
danger
Son you wanna be an Airborne Ranger

ROLLIN' ROLLIN' ROLLIN'

Rollin' rollin' rollin'
Ooh my head is swollen

Don't let your dog tags dangle in the dirt
Pick up your dog tags tuck 'em in your shirt

Rollin' rollin' rollin'
Ooh my neck is swollen

Don't let your dog tags dangle in the dirt
Pick up your dog tags tuck 'em in your shirt

Rollin' rollin' rollin'
Ooh my chest is swollen

Don't let your dog tags dangle in the dirt
Pick up your dog tags tuck 'em in your shirt

Rollin' rollin' rollin'
Ooh my hips is swollen

Don't let your dog tags dangle in the dirt
Pick up your dog tags tuck 'em in your shirt

UP JUMPED THE MONKEY

Up jumped the monkey from a coconut
grove
He's a mean mamma jamma you can tell by
his clothes

Rip stop cammies and I tan beret
This little monkey was here to play

Line a hundred sailors up against the wall
He bet a hundred dollars he could whip 'em
all

Whipped ninety-eight 'till his fists turned
blue
Then he switched his fists and he whipped
the other two

UP IN THE MORNING

Up in the morning too soon
I don't like it no way

I eat my breakfast too soon
Hungry as a hound dog before noon

I went to the mess sergeant on my knees
Said mess sergeant, mess sergeant feed me
please

The mess sergeant said with a big old grin
If you want to be Infantry you gotta be thin

Up in the morning too soon
Eat my breakfast too soon

I went to the mess sergeant on my knees
Said mess sergeant, mess sergeant feed me
please

The mess sergeant said with a big old grin
If you want to be Airborne you gotta be
thin

Up in the morning too soon
I don't like it no way

I eat my breakfast too soon
Hungry as a hound dog before noon

I went to the mess sergeant on my knees
Said mess sergeant, mess sergeant feed me
please

The mess sergeant said with a big old grin
If you want to be Ranger you gotta be thin

Dunnigan Industries, Inc.

HEY LOTTY DOTTY

Left, left, left, your right left
Left, left, keep it in step

Oooh right a left
Left your right a left

Oooh right a left
I love to double time
We do it all the time

Hey lotty dotty
(chorus) Hey, hey

Hey lotty dotty hey
(chorus) Hey, hey

Dress it right and cover down
(chorus) Hey, hey

Forty inches all around
(chorus) Hey, hey

Hey lotty dotty
(chorus) Hey, hey

Nine in front and six to the rear
(chorus) Hey, hey

That's the way we do it here
(chorus) Hey, hey

Hey lotty dotty
(chorus) Hey, hey

Standing tall and looking good
(chorus) Hey, hey

We aughtta be in Hollywood
(chorus) Hey, hey

RUN, RUN, RUN

My grandpa is 95
He still runs and he's still alive

My grandma is 92
She likes to run and sing some too

I don't know what I've been told
If you never stop running
You'll never grow old

Contributed by,
C/SrA Matt Bumgardner

YANKEES ON THE ROAD

92 yankees on the road
Running to the depot to get another load

One, two, three, four, hey
92 yankees on the way

Gotta get there by C.O.B.
I've got soldiers depending on me

One, two, three, four, hey
92 yankees on the way

Been to the depot, gotta get back
My 1SG won't give me no slack

One, two, three, four, hey
92 yankees on the way

Contributed by,
SFC Tracy Stephens

SEE YOUR CADENCE IN PRINT

We are always seeking new material for our cadence series. If you have a marching or running song you would like to contribute to future printings, please send them for review. Identify it as a running or marching cadence and print your name clearly for publishing recognition. Include your email address as a point of contact. Submissions can be sent to the P.O. Box address listed on the title page or email them directly to:

Leaders@DunniganIndustries.com

Dunnigan Industries, Inc. reserves the right to not publish cadences found to be distasteful or inappropriate. Those selected for publishing will be notified by email of when their cadence will be in print. Dunnigan Industries, Inc. offers no other compensation for contributed cadences other than personal recognition after the last line of selected cadence submissions.